PRINTING PLANT
BY MELVIN BERGER

INDUSTRY AT WORK

FRANKLIN WATTS · NEW YORK · LONDON · 1978

Photographs courtesy of:

3M Company: pp. 3 (top), 4, 8, 12 (right), 26, 30, 35, 37, 44, 48 (bottom); Jamestown (N.Y.) Post-Journal: pp. 3 (bottom), 23, 43, 51 (photos by M. Berger); American Book Company / Stratford Press: pp. 5, 56, 59, 60 (photos by M. Berger); Printing for Industry: p. 12 (left: photo by M. Berger); Pat De Meo, principal, New York School of Printing: pp. 16, 21, 42, 48 (top: photos by M. Berger); Newsday (reprinted by permission): p. 52.

Cover design by Charles Goslin.
Cover photographs courtesy of 3M Company.

Library of Congress Cataloging in Publication Data

Berger, Melvin.
    Printing plant.

    (Industry at work)
    Bibliography: p.
    Includes index.
    SUMMARY: Describes various steps in the preparation and printing of materials including books, magazines, and newspapers and introduces various jobs of people involved in this work.
    1. Printing, Practical—Juvenile literature. 2. Printing industry—Juvenile literature. [1. Printing, Practical. 2. Printing, Practical—Vocational guidance. 3. Vocational guidance] I. Title. II. Series: Industry at work series.
Z244.B563   686.2   78–2529
ISBN 0–531–02207–2

Copyright © 1978 by Melvin Berger
All rights reserved
Printed in the United States of America
6 5 4 3 2 1

CONTENTS

INTRODUCTION — 1

THE FRONT OFFICE — 7
Customer and Printer Get Together

THE COMPOSING ROOM — 14
Text into Type

PLATEMAKING — 25
From Image to Metal Plate

MAKEUP, PASTE-UP, AND STRIPPING — 40
Putting the Page Together

PRESSROOM — 46
From Plate to Paper

THE BINDERY — 54
From Pages to Books

FURTHER READING — 62

INDEX — 64

THIS BOOK IS DEDICATED WITH DEEP AFFECTION TO HERMAN GOODRICH, WHOSE COLONIAL PRESS WAS MY INTRODUCTION TO THE FASCINATING WORLD OF PRINTING.

ACKNOWLEDGMENTS

I am most grateful to the many people in the printing industry who helped me with the writing of this book. My particular gratitude goes to Pat De Meo, principal, Patrick J. Delaney, assistant principal, and the staff and students of the New York School of Printing; to David D. Leddy and Harry Wolff of the American Book Company/Stratford Press; to Sidney A. Sweeney of the Jamestown (N.Y.) *Post-Journal;* to Ralph Silverman of Printing for Industry; to James C. Radford of the 3-M Company; to Joanne C. Danaher of *Newsday;* and to John D. Young of the Printing Industries of America.

INTRODUCTION

Printing is one of the largest industries in America today. The need for the printed word has never been greater. Just about every one of us uses many forms of printing each day. We get information from printed books, magazines, and newspapers. We identify products by their printed labels. Businesses sell through printed advertisements. We spend printed money, and play with printed games. We use printed catalogs and maps and hang printed posters on our walls. And we even wear T-shirts with printed words or pictures on them.

Almost every city and town in the country has a printing plant. In fact, there are more printing plants, ranging in size from storefront shops to immense factories, than any other single kind of industrial plant. Every year these plants produce printed products that are worth more than $25 billion. Night and day, the printers are working and the presses are rolling in these plants. They are turning out the vast amounts of printed material that are needed by our culture and our economy.

All printing plants are places where many identical copies are made of certain images. These images may be letters and words, or pictures, or both. Most printing is done on paper. But printing can also be done on cloth, metal, wood, and plastic surfaces.

Over one million men and women work in printing plants. About 500,000 of these people work for job printers or commercial printers. These shops do jobs varying in size from a run of 50 letterheads to a run of a million pamphlets. Some 350,000 men and women print the newspapers that are published every day or every week across the nation. Book publishers employ 100,000 printers. They produce all the books that you find in bookstores and libraries. Another 75,000 printers are at work producing the many different magazines that people read.

Some printers are highly skilled craftspeople. They are in charge of the big, expensive machines and presses, and are responsible for one or more operations in the plant. Most skilled workers learned the basics of their trade in school, and then received more training by working as apprentices in printing plants.

Some printers are semiskilled. Their jobs entail less responsibility. Often they feed the machines or presses, or perhaps operate some of the simpler machines. They are usually taught their tasks by more experienced workers in the plant.

There are also many unskilled workers in printing plants. They work as laborers or assistants in jobs that

Above: printers are at work in the printing plants found in every city and town in the country. Right: some 350,000 printers work for the country's newspapers.

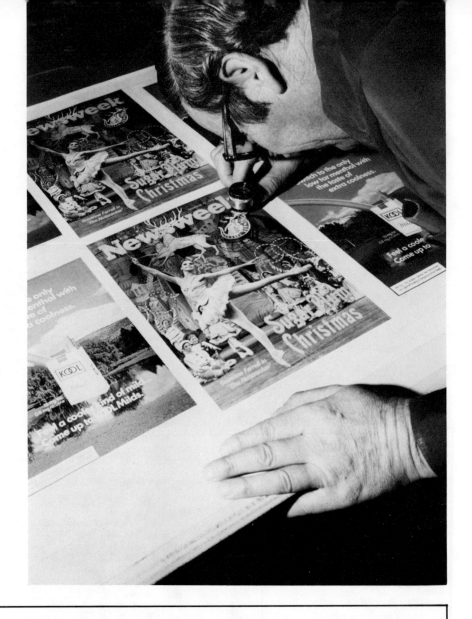

One of the 75,000 printers who produce magazines studies a proof of a magazine cover through his magnifying glass.

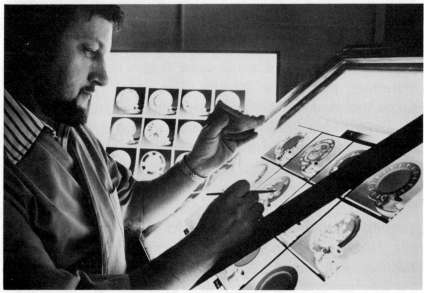

Above: there are 100,000 workers who are employed by book publishers. Below: many skilled artists perform varied tasks for printing plants.

require little or no training. They always serve under the direction of more skilled printers.

Besides the men and women who actually do the printing, many other people work in printing plants. These include secretaries, clerks, artists, designers, mechanics, and electricians.

Just now the printing industry is in the midst of a revolution. Big changes are taking place. More and more computers, electronic devices, cathode-ray tubes, laser beams, special cameras, and other highly advanced machines are being used in printing plants today. They are making printing easier, faster, and more attractive-looking than ever before.

But printers do not just want to turn out more printed materials at greater speeds. Too many printed messages just end up in the wastebasket without ever being read. To avoid printing unappealing material, the printers are also learning how to make the messages more attractive and pleasing. The task of the modern printer is to use lines, shapes, colors, and textures to improve the effect of a printed message. This advance is called graphic communication. The word graphic refers to anything that is written or drawn; communication is giving information to others. Graphic communication, then, joins the mechanics of printing with the art of communication.

Graphic communication is a field based on the long, rich history of printing. The challenge of using print to attract the reader's attention, to bring people knowledge and understanding has never been more important.

THE FRONT OFFICE
CUSTOMER AND PRINTER GET TOGETHER

The front office of a printing plant looks much like any other office. There are chairs, desks, and bookshelves. The only difference is that the front office of a printing plant has stacks of printing products catalogs standing on the shelves or open on the desks.

There are catalogs that show the many different grades, colors, textures, and weights of paper. Samples of each type of paper are contained in big, thick volumes.

There are ink catalogs that show all of the shades, colors, and thicknesses of the inks that are available.

There are sample books of typefaces, with letters of the alphabet and numbers shown in a great variety of styles and sizes.

Finally, there are catalogs of printing equipment, with pictures and descriptions of the latest machines that are on the market.

The sales representative asks for the production worker's suggestion on the best way to do the job.

PRINTER'S REPRESENTATIVE

Most large printing plants employ workers, called printers' representatives, to help customers choose the best paper, ink, type, and equipment for their job. Printers' representatives, called reps for short, are really salespeople. They are the go-betweens for the printing plant and the customer. The chief requirements for this job are a complete knowledge of the printing industry and a pleasant personality.

The reps' desks are in the printing plant's front office. They are separate from the production department, where the actual printing is done. Either they meet with the customer in this office, or they visit the customer's offices.

If the customer wants to know how he or she can communicate through a printing job, the rep points out what the plant can offer. Together they go through sample charts, picking the best paper, ink, and typeface for the job. They arrive at a rough idea of what the finished article will look like and what it will cost.

Most customers, though, before they place an order, want a cost estimate of the job. For a complete breakdown of cost, the reps usually turn to the estimators who work with them in the front office.

ESTIMATOR

The estimator's job in most large printing plants is to prepare accurate figures on the cost of each printing job.

They also find ways to do the job at the lowest cost to the customer, and at the lowest cost to the printer.

Besides catalogs and sample books, estimators keep calculators or adding machines on their desks. Since all of their work deals with figures, they depend on these automatic machines for speed and accuracy.

The estimators compute the cost of the supplies and materials the customer's job will need. What will it cost for the paper, ink, and metal plates that cannot be used again? To this number the estimators add the labor costs. How many hours will be spent preparing the type or plates? How many hours to set up and run the job on the press? How many hours of cutting, folding, or stapling the printed sheets? Finally, the estimators add a part of the general cost of operating the plant. They include a portion of the costs of the machines, rent, heat, light, taxes, salaries of those not directly involved with the job, and a fair margin of profit for the owner of the plant. The total is the cost estimate for the job.

The estimators usually come to their jobs after years of working in the production end of the printing industry. They must know printing well enough to make very accurate guesses of the cost of each step of the printing process. At the end, the charges must be high enough to give the company a fair profit, yet low enough to win the customer's approval.

LAYOUT ARTIST

The rep also calls on the plant's layout artist to help prepare a drawing of the printing job. The layout artist is able to visualize what the rep and customer have in mind. Working at a large, flat, tilted drawing table, he or she makes several very rough sketches at first. How much space for text and how much for illustration? What style of lettering will look best and carry the message in the clearest way? The artist letters the text in freehand to see how it looks with the illustrations. He or she may use Magic Markers to experiment with color and design. When everything is worked out on the roughs, the artist prepares a final, full-sized sketch, called a dummy.

Most layout artists studied commercial art either in college or in special art schools. To be successful, they must be able to draw well and quickly, and have a very good sense of design, form, and color. They must also be familiar with the many different kinds of type that are available, and they must know enough about printing to know what is, and what is not, possible.

The rep presents the estimator's figures and the artist's dummy to the customer. If they are satisfactory, a contract is signed.

The rep will now follow the job all the way through the plant until it is finished. At each step, the rep will get the customer's approval before going on to the next step. The customer tells the rep if anything is wrong, and the rep then speaks to the people in charge of production, who correct the mistake.

Probably the most important worker "up front" is the chief executive. This is the person who directs the operation of the entire plant, and is responsible for all final decisions. Other executives may be in charge of the various departments, such as production, sales, art, per-

Left: the layout artist prepares a sketch to show where the text and the illustrations will be placed on the page. Right: the sales representative follows the customer's job through the plant. Here the printer points out a detail on the printed page.

sonnel, and so on. But the chief oversees the operation of all parts of the plant.

In small plants, one person, often the owner, will be printer's rep, estimator, and executive. And that person may also do the printing job once the contract is signed.

But, whether it is very large or very small, the work in the front office is the first link in a chain that stretches from the customer's original idea to the final printed material.

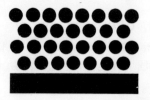

THE COMPOSING ROOM
TEXT INTO TYPE

Once the customer signs the contract, the printing process is ready to begin. The first step is to change the customer's words, called copy, into a metal form. This procedure is known as setting type, or typesetting. The workers are called either typesetters or compositors. They are found in the composing room of large printing plants, or in separate plants that only do typesetting.

HAND SETTING

There are several different methods of setting type. The oldest method is to set type by hand. Although most typesetting today is done by machine, there are still some special printing jobs that are best set by hand. Machines, for example, cannot always set type that is very large or unusual in shape, such as that used for some advertisements. And for small printing jobs, like putting someone's name on a hundred Christmas cards or napkins, it is easier and more efficient to set the type by hand.

For these hand-set jobs, the typesetters spell out the words by selecting separate pieces of type from a large, flat box with small compartments, called a case. Each piece of type is a small metal bar, with a single letter raised a fraction of an inch above the surface of one of the ends.

The typesetters hold a stick, a small metal frame, in their left hand. They pick up the individual pieces of type with their right hand, and line them up in the stick.

As soon as the typesetters fill up their sticks with about two inches of type, they set the type into a metal frame, called a galley. When the galley is filled, they lock the type into place to keep it from falling out.

The typesetters then spread ink over the type with a roller. They place a sheet of paper over the inked type and, using a press, force the paper down tightly over the letters. The ink on the letters is transferred to the paper. They carefully read this proof, as it is called, for errors. If the typesetters find any incorrect letters, they exchange them for the correct ones.

Hand-set printing makes it possible to use the same pieces of type over and over again for different printing jobs. But it is a slow and laborious process. Picking out the individual letters takes time. Then, after the job is done, replacing them in the proper compartments of the case is also very time consuming. In printing, whatever takes a lot of time, costs a lot of money. And there is

The student learning printing in high school
holds the stick in her left hand
as she picks the type with her right hand.

always the danger that the compositor will "pi the type," that is, drop all the letters out of the stick. For these reasons, almost all typesetting is done by machine.

HOT METAL: LINOTYPE

One of the most popular typesetting machines is the linotype. It is a big, ungainly looking machine. It makes a lot of noise when in use, and often has an acrid, burning smell around it. Yet it is the fastest, most dependable, and most frequently used of the hot-metal typesetting machines.

What looks like a typewriter keyboard, with nearly a hundred keys, extends out from the front of the linotype. The linotype operator sits at this keyboard. Above the keyboard is the copy that is to be set. A small, dim bulb casts a light over the front of the dark, hulking machine.

The linotype operator types out the letters and words of the copy. Each time the operator presses a key, a hollow brass mold called a mat, of that letter slips down into the assembly box inside the machine. As soon as the letter molds for an entire line of type are in the assembly, the operator presses a special lever. Molten metal within the machine is forced into the hollow molds.

In a few minutes the metal cools off and hardens. It forms one continuous, line-long piece of metal, with the letters raised along one edge. It looks like a line of type that has been set with separate type, except that they are all joined together.

The name, linotype, comes from the fact that it forms a single line of type. And it is called a hot-metal machine because the metal, a mixture of lead, tin, and antimony, is heated to the melting point so that it can be poured into the molds.

The finished line of type is called a slug. Each slug slides out into a tray on the machine. But the linotype operators do not wait for the slug to appear before they begin typing the next line. As soon as slugs for an entire page are ready, an assistant places them in a frame, locks them in place, and pulls a proof to check their accuracy.

Linotype operators are skilled workers. Many of them learned their trade through apprenticeships that may have lasted as long as six years. Some learned how to run the machine at printing school, and then served as apprentices for about two years.

Running a linotype is a demanding job. The operators must concentrate on the copy, no matter how dull or uninteresting it seems to them. They must always set in type exactly what the copy says, and never what they think it should say. And they must do it very fast and very accurately.

A good number of printing plants have other hot-metal typesetting machines similar to the linotype. One of the more popular is the monotype. As the operators type in the copy, the letters are changed into a code of round holes punched into a long, narrow paper tape.

When the operator has completed the punched

tape for an entire page, another composing room worker feeds it into a separate casting machine. The casting machine then produces the slugs of type.

COLD TYPE: PHOTOTYPESETTING

In many printing plants a new photographic method of setting type has been introduced. It uses neither the separate pieces of type of the hand-setting method, nor the slugs of the hot-metal machines. In fact, it does not use metal at all. Instead, it actually uses photographic images of the letters. This approach, so different from the others, is called cold type.

The operators of phototypesetting machines sit at typewriter keyboards, similar to those on the linotype machines. Each time they strike a key, though, an image of that single letter is flashed on to a strip of photographic paper inside the attached machine. At the same time the letter also appears on a TV screen, called a Video Display Terminal (VDT), over the keyboard. This allows the operators to check the accuracy of their work as they go along.

When the operators finish a complete article or page, the photographic paper must be developed to make the words visible. Sometimes they bring the exposed paper to a separate machine where it has to go through several chemical baths in order for a photographic print of the words to be made. In the more modern phototypesetting machines, the film is automatically developed within the machine. The finished print looks exactly like a single printed copy of the original text.

While it is often dirty and grimy around the linotype machines, the phototypesetting machines are usually located in light, clean, attractive areas. These operators do not need to wear the overalls or aprons worn by linotypists because they do not handle ink-stained type or dirty metal slugs. Very often they just wear their regular street clothes.

Very modern printing plants have much improved phototypesetters. In one machine, instead of projecting an image of the letter onto photographic paper, the letter flashes on a TV screen to expose the photographic paper. In another, as the operator types the letters, a punched paper tape is produced similar to that made by the monotype method. The operator later runs the tape through a phototypesetter that makes the photographic print of the type. A good number of the most advanced phototypesetters have built-in computers that do many of the tasks that had been done by the human typesetters.

The operators of phototypesetters must be specially trained for their work. They must learn how to use the machine, and to understand what happens when they hit each of the keys on the keyboard. They must have a knowledge of photography, electronics, and computer science so that they can adjust the machine to work properly, and make minor repairs. And they must be good typists, able to type quickly and accurately.

When the phototypesetting is finished, no matter

For cold type or phototypesetting the operator sits at a keyboard that looks like an electric typewriter.

which particular machine is used, the result is a sharp, clear photographic print of the page.

TYPESETTING FOR NEWSPAPERS

The typesetters who work in newspaper printing plants use many of the same methods found in other parts of the printing industry. But since they need to set so much copy each day, and have to work with great speed, they also have some special approaches.

Stories from news services, such as Associated Press and United Press International, are sent to newspapers all over the country. They are sent as a code of electrical impulses at very high speed over the regular telephone wires. They are received at the newspaper offices on special teletype machines. The electrical impulses automatically cause the teletype machines in the newspaper offices to type the story on a long strip of paper.

From time to time, an editor tears off a length of paper, and decides which stories to use. These stories are given to the typesetter, who places them in a computerized machine, called a scanner or automatic reader. Using a laser beam, or some other method, the machine automatically reads the typewritten material from the teletype machine.

The operator reads the material that appears on the VDT screen above the machine. To make any changes or corrections, the operator strikes various keys on the attached keyboard. When satisfied, the operator hits a

Some of the stories arrive at newspaper offices in the form of punched paper tape. The printer runs the tape through a machine that produces cold type composition.

key that places the entire story in the memory of a computer. Later, it will be used to produce a photographic print of the story.

In some newspaper offices, the same stories arrive as punched paper tapes, as well as in typewritten form. Again, the editor chooses the stories to be printed. A worker then runs the tape, often called an "idiot tape," through the phototypesetter to produce the photographic print.

When the typesetter is finished, the copy must be proofread, regardless of whether it is hand set or machine set, hot-metal type or cold type. Special proofreaders check every word as carefully as they can to make sure that there are no mistakes, which are called typos. If they find anything wrong, the copy goes back to the typesetter to be corrected or done over again. Finally, the printing job is ready to move on to the platemaking part of the printing plant.

PLATEMAKING
FROM IMAGE TO METAL PLATE

Most printing today is done from metal plates, rather than directly from type. One reason is that type is made of soft metal that wears down when many copies are made. Also, since all illustrations must be printed from plates, it is easier to use a plate when printing both text and illustrations.

Platemakers transfer pictures and typeset copy onto metal plates. They are highly skilled workers who spend many years mastering their craft. Either they work in one section of the printing plant, or in separate factories that only make plates.

PLATES FROM PICTURES: LINECUTS

When a job includes a black-and-white picture, such as a drawing, map, chart, or cartoon, it is given first to

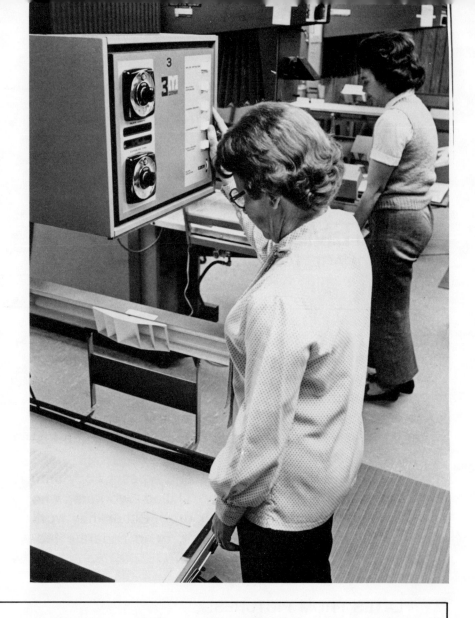

The first step in making a printing plate is for the camera operators to photograph the copy.

the camera operators. These well-trained workers are in charge of the large-size cameras that are found in every platemaking shop.

The camera operators place the picture in a frame that is attached to the camera. They make sure that the picture is held flat and is securely attached. They adjust the camera to be sure that they will get as clear and sharp a photograph as possible.

The camera operators then turn on several blinding lights that are aimed at the picture. In a few seconds the lights automatically go off.

The next step is to develop the exposed film. Most platemaking plants now have film processors that develop the film automatically. When the film comes out of the processor, it is in the form of a negative.

A negative looks like the opposite of the original picture. Suppose, for example, that the original picture is a black square on a white background. Then the negative is a transparent white square on an opaque black background.

The camera operators give the finished negative to the photoengravers. These are men and women who make printing plates by photographic means.

The photoengravers place the negative on a thin metal plate. The plate, usually made of either zinc or copper, has been coated with chemicals that harden under light.

The photoengravers then shine a powerful light through the negative onto the plate. Since the image (say of the black square) is transparent on the negative, the light passes through that area and strikes the plate. The light changes the chemicals on the plate, which form a solid protective coat on the metal where the light has struck. Since the background part of the negative is dark, no light gets through. The chemical coating there remains soft.

Next the photoengravers put the plate through several chemical steps, including a bath in a very strong acid. The soft part of the chemical coating is washed off. It allows the acid to eat away the metal of the plate where it is not protected by the hard coating. At the end, the image (the square) remains raised up over the rest of the surface.

The photoengravers carefully examine the plate, now called a linecut. They look to be sure that the raised image has good, clean edges. Quite often they use a drill-like tool to grind away parts of the background that have not been eaten away deeply enough. They also use other tools to fix up any further defects in the plate.

To get the plate ready for printing on the press, the photoengravers glue or nail the plate to a block of wood. This makes it the same height as a raised letter of type.

Before passing the finished linecut onto the press operators, the photoengravers ink the plate and pull a few proofs. They examine the proofs very carefully for any imperfections. When they are satisfied with the results, the plate is ready for the next step.

HALFTONE PLATES

Linecuts can only be used for black-and-white pictures. They cannot be used to make plates of photographs or other illustrations that include shades of gray. To print grays, the photoengravers need to make halftone plates.

For halftone plates the camera operators insert a special glass screen inside the camera. The screen is covered with thin, crisscrossing lines. There can be anywhere from 55 to 200 vertical and horizontal lines to a square inch.

The camera operators photograph the picture just as for a linecut. The screen, though, breaks the image into tiny dots. Where the picture is black, the dots are large, making the image appear almost solidly black. In the gray areas, the dots are smaller and seem farther apart. Where the original picture is white or nearly white, the dots are still smaller and look still farther apart.

If you look closely at almost any printed photograph, you will see the pattern of dots. You can see the pattern of dots most clearly in newspapers, which use screens having 60 to 75 vertical and horizontal lines to a square inch. It is somewhat harder to see the dots in this book, which used a screen having 133 lines per square inch. The greater the number of lines, the harder it is to see the dots, and the better the reproduction of the original picture. In all cases, the dots can be seen up close. But at a distance, the individual dots disappear, and all you see are shades from white to light gray to dark gray to solid black.

The camera operators develop the film automatically and give the negatives to the photoengravers. The photoengravers place the negative on a coated metal plate and shine a light through the negative. They then

In most plants the making of the plates is an automatic process. This platemaker is guiding the finished offset plate as it comes out of the platemaking machine.

add chemicals and acids to eat away the plate. Thus far they have done just what they do in making linecuts. But instead of raised surfaces on the plate, there are raised dots: some larger and closer together, some smaller and farther apart.

Photoengravers often have to touch up their halftone plates. If an area is too dark, they protect the rest of the plate with an acid-resistant chemical, and apply acid to the dark section. This makes the dots smaller, and the area becomes lighter.

If an area is too light, the photoengravers burnish the plate. They rub over the dots with a metal tool that flattens and spreads the dots. The larger the surface area of the dots, the darker the halftone when it is printed.

As with linecuts, the photoengravers pull several proofs before sending the halftone plate on to the next step.

DUPLICATE PLATES

Besides making plates of pictures, platemakers are often asked to make plates, called duplicates, from type. Duplicates are used for long press runs. They are also used when big print jobs are run off at more than one plant, or when more than one copy is printed at a time. And they are needed for high-speed rotary presses, as opposed to flat presses, where the printing surface must be curved to fit onto a large cylinder.

A stereotype is one of the simplest, cheapest, and fastest ways to make a duplicate plate. The worker first places a thick sheet of cardboard-like paper, called a flong or matrix, over the metal type and places it in a press. Under pressure, the soft matrix takes on the exact shapes of the type, making a mold that is a faithful copy of the letters and words.

In a casting box, the platemaker then pours molten metal into the mold, creating a duplicate printing surface. For a rotary press, the mold is curved in the casting box, so that the stereotype plate is also curved.

Many newspaper printing plants use stereotype plates. Magazine printers, though, prefer to use the more exact electrotypes.

For electrotypes, platemakers use a press to make a mold of the type in a thin sheet of lead or plastic. Next they hang the mold in a chemical solution, along with a copper bar. Then they flip a switch that sends an electrical current through the solution. This allows a thin copper coating to form on the inside of the mold. After removing the copper shell, they back it up with lead to make it strong enough to be used for printing.

Platemakers use the actual metal type for making stereotypes and electrotypes. How do they make duplicate plates from the paper or film of cold type composition?

They go through almost exactly the same steps as when they make a linecut: Take a photograph of the cold type composition, make a negative, expose a coated plate through the negative, and apply acid to eat away the exposed areas. The result is a metal plate with the letters raised above the rest of the surface.

This approach is also used to make a plate of copy set in metal type. A worker pulls a single, carefully made proof of the type. Since it will be used to reproduce the type in a plate, it is called a reproduction proof, or repro, for short. The platemaker then uses the repro in the same way as the cold type composition to prepare a plate.

There is now new equipment to make plates from cold type composition or repros much more easily. The platemaker merely places the copy in a scanning machine. Inside the machine, laser beams read the copy, and then automatically cut a plate.

OFFSET PLATES

Photoengravings and duplicate plates both have raised surfaces to transfer the images to paper. But there is another way to print. In this approach, absolutely flat plates are chemically treated so that ink sticks to the images of letters and pictures, but not to the rest of the plate.

The flat-plate method is called photo-offset lithography. Usually the name is shortened to offset. More and more printing plants are now using offset. It is rapidly replacing the older method, which is called letterpress.

All offset printing is done from plates. The process of making the offset plates is similar to that of making photoengravings.

Camera operators photograph the cold type composition, repros, or an original picture. They develop the film and produce a negative.

The platemakers place the negative on a plate in the glass frame of an automatic platemaking machine. The plate is usually a thin sheet of aluminum that has been coated with a light-sensitive chemical. Since the chemical is also sensitive to the regular light in a room, the offset plates are made in rooms equipped with special yellow lights that do not affect the chemical.

The platemakers lock the negative and plate in the platemaking machine. They adjust the controls so bright lights go on inside the platemaker, which transfers the image to the coating on the plate.

The platemakers then put the plate through several chemical steps, either automatically or by hand. When it comes out, only the coating of the image will accept and hold ink. The coating over the nonprinting areas will only accept water. Special greasy ink is used for offset printing, and water is spread on the nonprinting areas. Since grease and water do not mix, offset printing is able to produce sharp, clean images.

As soon as the plate is checked, and the platemakers pull proofs, the offset plate is ready to go to the press operators.

GRAVURE PLATES

There is still another basic way to print. In gravure, or intaglio printing, the image is made up of tiny wells that are cut below the plate's printing surface. A number of

How did it come out? The platemaker compares the offset plate with the original copy.

Sunday newspaper magazine sections and some entire magazines are printed from gravure plates.

To make a gravure plate, the camera operator first photographs the copy, words as well as pictures. From this point on, the steps are the opposite of those for making halftone photoengravings.

Instead of a negative, the operator produces a positive. When the platemaker exposes the new plate to light, the light is not able to pass through the area of the image.

In the acid bath the chemicals eat into the plate in that area. Where the original was black, the acid creates large, deep wells in the surface of the plate. Where the original was gray, the wells are smaller and shallower. And in the light areas, the wells are smaller yet and even more shallow.

When gravure plates are printed, ink is sprayed into the wells, and any ink on the well walls is wiped off. By using a great deal of pressure, the paper picks up the ink from the wells, transferring the image to the paper.

COLOR PLATES

All color printing is done from plates. In recent years, new machines have improved the quality of the plates so that more and more printing is now done in color.

The simplest way to print color is by the flat-color method. It is used when the colors are printed without shading, such as in the Sunday comics.

The platemakers prepare flat-color plates as they do

Color printing is very difficult. The printers must be very careful to duplicate the exact colors of the original.

linecuts, but with some differences. Suppose they want to print a picture with red, yellow, and black. The camera operators will then take three full-size photographs of the picture, using black and white film. They will develop the film producing three identical negatives.

On one negative, a retouch artist paints over, or opaques, all areas except the area that will be printed yellow. On the second negative, the retoucher opaques everything but the red. And on the third, the retoucher opaques everything but the black.

The three negatives then go to the photoengravers. They use them to make plates in the same way as they make ordinary linecuts. None of these plates, though, shows the entire picture. One shows the parts that will be printed yellow, the other shows only the red parts, and the third shows only the black parts.

Later when they are placed on the presses, the press operators will use yellow ink when they print the first plate. Then they will run the same paper through again, using the red plate and red ink. For the third run, they will use the third plate and black ink. Most large printing plants have large presses that print all the colors, one after the other, on the same sheets of papers.

But what if the picture has many different shades and hues of color? Then the printer must use a more complicated method, called full-color or process-color printing.

The first step is to separate all the primary colors in the picture and make a plate for each one. The camera operators start this process by photographing the picture with a colored filter in the camera. A filter allows light of only one color to pass through the film. For example, to separate yellow, the operators use a violet filter. Violet is made up of red and blue. The violet filter stops the red and blue light, and only the yellow gets through.

In the same way, they use a green (yellow and blue) filter to get red, and an orange (yellow and red) filter to get blue. To separate the black, they use a special filter that blocks all three of the primary colors.

The camera operators develop the film producing separate yellow, red, blue, and black negatives. They then make a color print of each negative. They photograph each negative through a halftone screen to make a screened color negative. They turn the screen slightly, though, for each one, so that the dots do not come out on top of each other.

Photoengravers then use the screened colored negatives to make separate halftone plates in the ordinary way.

There is now new advanced electronic equipment that eliminates the photographic steps in full-color printing. The picture is placed in a machine where it is automatically scanned and each of the colors separated. It works very fast and accurately, and often produces better results than the older method. More and more platemakers are switching over to this approach.

Before passing the full-color plates on to the press operators, the platemakers pull a series of proofs, called progressives. They pull a proof of each color and in combination with other colors. In this way they can be sure that each of the plates is good, and that the final printing will be a faithful reproduction of the original artwork.

MAKEUP, PASTE-UP, AND STRIPPING
PUTTING THE PAGE TOGETHER

For simple printing jobs, the typeset copy goes from the typesetters to the platemakers. But for complicated jobs, with pictures, different kinds of type, or several columns of type, it must first go to the makeup or paste-up workers. Their job is to bring together all the separate elements, and arrange them into complete pages. Makeup workers do it with metal type and engravings. Paste-up workers do it with cold type composition and repros.

MAKEUP AND PASTE-UP WORKERS

Makeup and paste-up people follow the directions in dummies that have been prepared by the layout artists. These sketches are guides to the approximate location of each element on the page.

Makeup workers usually stand at flat metal tables,

wearing full-length aprons. They collect all the hand-set and machine-set type, and the photoengravings. They makeup, or arrange, them to form complete pages. Then they securely lock all the material for each page into a full-size sturdy steel frame, called a chase. The chase is then used to produce a duplicate plate. Sometimes it is used directly for printing.

Paste-up workers create complete pages by pasting pieces of paper onto large sheets of smooth, white cardboard. The pieces of paper include cold type composition, repros, and black-and-white pictures. (Pictures requiring halftone are handled separately.) They never touch a piece of type or a chase.

Most paste-up people sit on high stools at large, tilted wooden tables. They wear regular street clothes. Their tools are a large can of rubber cement for pasting, and a T square for lining up the elements to be pasted. They assemble all of the elements on the cardboard. As soon as everything is in place, they spread rubber cement on the backs of the papers and paste them into place.

Paste-up workers in newspaper printing plants spread wax on the copy pieces that they are arranging. The wax holds the paper on the cardboard, but not very tightly. If another news story comes along, the item can easily be removed or shifted elsewhere.

The hardest part of paste-up work is being sure that every section of text and every picture is lined up absolutely straight, and is evenly spaced.

The paste-up men and women are usually trained in drafting or mechanical drawing, which gives them the

The makeup worker begins to place the type for the text into a metal frame, called a chase.

The paste-up worker at the newspaper plant attaches the articles, pictures, and ads for an entire page onto a large piece of white cardboard.

The stripper carefully cuts windows in the goldenrod, exposing the negatives.

necessary skills. They often use light blue lines on the cardboard as guides. Sometimes the blue lines are already ruled on the cardboard, as on newspaper pages, which are always the same size. More often the workers have to draw the lines themselves. Light blue is used because the workers can see it, but it is invisible to the camera.

STRIPPERS

Camera operators take pictures of the finished paste-ups, and of any halftone illustrations for that page. They give the negatives to the strippers.

The strippers tape, or strip, the negatives onto pieces of heavy, ruled orange paper, called goldenrod. They very carefully position the paste-up and halftone negatives in place. Then they use a steel ruler and a sharp knife to cut openings in the goldenrod. The openings, called windows, expose the parts of the negative that are to be printed.

The goldenrod, with negatives in place and windows cut, is called a flat.

The strippers send the finished flats to the platemakers. The platemakers use them to make plates in the same way that they use ordinary negatives to make plates for simpler printing jobs.

PRESSROOM
FROM PLATE TO PAPER

The actual printing is done by printing press operators in the pressroom. The noisy presses are usually kept in a separate room, away from the other workers.

There are presses so small that they are kept on tables; others fill entire large rooms. Some are shiny and new; others are as much as fifty years old. Some are used for short runs of small jobs; others to print hundreds of thousands of copies of magazines or newspapers.

But basically, all presses work in the same way. Ink is spread over the printing surface, whether it be type, engraving, or plate. The press brings the paper against the inked surface, and the image is transferred to the paper.

MAKEREADY

Before starting to run the presses, the press operators have to get all the supplies and materials together and be sure that everything is set for the run. This is called makeready.

The press operators follow the instructions on the order form that tell which paper and ink to use. They also study carefully the finished printing plate or the form with the type locked in place. Is the surface undamaged? Is everything right for printing?

Next, the press operators place the plate or form in the press. There are three basic types of presses. In each one the plate or form is attached in a different way.

The platen press is the oldest, simplest, slowest, smallest, and yet one of the most popular of all presses. It is used only for short runs of small pages, such as wedding announcements, personalized Christmas cards, business forms, letterheads, and so on.

Printers usually print directly from type that is locked in a form. They attach the form in an upright or vertical position. Facing it, like the other half of a giant clamshell, is a flat surface, called the platen. The platen is hinged to the form side on the bottom.

Paper is placed on the platen surface. The ink comes from rollers that first pass over a large, ink-covered disk above the form, and then over the type. Then the platen, with the paper on it, is swung forward and up to press against the form.

The cylinder press is faster than the platen press. It uses either a form or a plate, which the printers lock into a horizontal position. When the press is running, the form moves back and forth under a series of rollers. First, rollers spread ink on the printing surface. Then the impres-

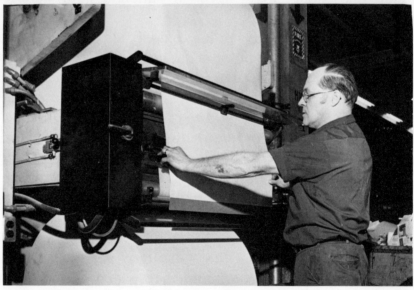

Above: the worker sets the form into the platen press. He will later ink the dish over the form, and the paper will be fed to the platen. Below: the printer threads the ends of the rolls of paper into the large rotary press.

sion cylinder grips a blank sheet of paper, and presses it down to receive the printed image.

The rotary press is the fastest of all presses. Several rotary presses can be combined to do big jobs, such as those required by big-city newspapers. A long line of combined rotary presses can produce up to a million complete newspapers an hour!

Rotary presses only print from curved plates, which the press operators attach to large cylinders. For letterpress, the curved plates print directly on the paper. The plates in the offset rotary presses print on a rubber roller, which then prints on the paper.

While the press operators are placing the form or plate on the press, other workers are preparing the paper for the pressrun. Workers on the small presses are placing supplies of ready-cut sheets on the press.

Workers on the larger and faster presses, though, do not always use ready-cut paper. Some of these presses print on big, long rolls of paper, called webs. For the web presses, the worker usually uses a mechanical hoist to get the heavy roll of paper into position. Then the worker threads the end of the paper in, around, over, and under all the rollers and cylinders in the press. After printing, the press will automatically cut the paper into sheets.

Once everything is set, the press operators start up the press. They let it run for a few minutes as they check whether it is running smoothly. Is the paper feeding correctly? Is the ink covering the printing surface? They even listen to make sure that the press motor and moving parts sound right.

Almost always they find something out of adjustment that must be fixed. Usually it is a minor problem, and the printers handle it by themselves. If it is a serious problem, they may have to call the plant's mechanics to make the repair.

The press operators also look at the copies that were printed during the short run. Is the printing in the right place on the page? Did any parts not print at all? Are any parts too dark or too light? Is there any dirt on the plate or type that shows up on the printed copies? Is there any smudging?

Again, there are always some adjustments the pressroom workers have to make. They change the pressure of the paper on the plate. They put some backing material behind one part of the plate or some of the type to make it higher. They shift the position of the plate. They change the flow of ink. They tighten the mechanism that feeds the paper.

Then, they start the press again. And once more they check and adjust until they are satisfied that they are getting the very finest impression. All this work, the preparation and makeready, may take longer than the actual pressrun.

RUNNING THE PRESS

Once everything is set, the press operators start their run. Almost all presses in use today are automatic. The operators merely press a button or throw a switch, and the presses start rolling.

While the presses are running, the printers watch carefully to see that everything is going smoothly.

The finished newspapers move on conveyor belts from the press to the loading platform.

The very large rotary presses are painfully loud when they run. The men and women who work around these presses often wear earphones to protect their hearing.

Even though the presses are automatic, the printers do not relax during the run. There is still plenty to do. They constantly check that the presses are running smoothly, that the copies are coming out right, and that there is a good supply of ink and paper in the machine.

As soon as the run is over, the operators usually take the finished job from the pressroom to the shipping room, where it will be packed and sent to the customers. If it is a newspaper, the printed and folded papers move on conveyor belts to a platform where they are loaded on trucks. If it is the pages for a book, the job is taken to the bindery.

The pressroom workers now remove the used plates or forms from the press. They wash the ink off the rollers and clean up the paper, dust, and ink stains from the press. And as soon as this work is done, they start the preparation and makeready for the next run on the press.

THE BINDERY
FROM PAGES TO BOOKS

Every book starts out as one or more very large printed sheets of paper. The printing plant's large presses can turn out thousands of printed sheets an hour. The printed sheet usually has thirty-two separate pages printed on each side of the paper.

The printed sheets are made into books at the bindery. Sometimes the bindery is one part of the printing plant. Sometimes it is a separate plant. The bindery workers cut, fold, gather, sew, glue, trim, and attach covers to the printed sheets to make the books.

FOLDING AND CUTTING
THE PRINTED SHEETS

The printed sheets arrive at the bindery in tall stacks on wooden skids. The first step in the bindery is to slit and fold the printed sheets. The folding machine operators lift a number of sheets by hand and load them on to the board of the folding machine.

Automatically the sheets are fed into the machine. They are slit into smaller sheets. These sheets are also folded several times so that they are about the size of the pages in a finished book. They are called signatures.

An assistant awaits the folded signatures as they come out of the machine. Stacks of them are piled one on top of another on a rack. With a flick of a switch, a weight comes down and presses the air out of the signatures, making them flatter and easier to handle. The stacks of signatures are strapped together and passed to the workers who will gather them together in the correct order for the book.

While the signatures are being slit and folded, the parts of the book not printed as part of the text are being cut to size. These include some color illustrations, and the endpapers, the heavy pages at the very beginning and end of the book. A worker at a cutting machine pushes the control button and a sharp blade quickly slices through the big stacks of paper.

The signatures and the other pages are all brought to the gathering section of the bindery. In one large bindery, the long, narrow gathering machine has forty-four pockets, or hoppers, to hold signatures. Some workers bring the signatures from the folding machine to shelves along the gathering machine. Others load the pockets of the gathering machine with the successive signatures that will make up the entire book.

A mechanical arm lifts a signature from the first

The bindery worker stacks the signatures in the hoppers of the gathering machine. Mechanical arms will pick them up in order and place them on a conveyor belt on the other side.

pocket and places it on a slow-moving conveyor belt. Another mechanical arm takes a signature from the second pocket, and puts it on top of the first signature. This process continues down the line until all the signatures for a particular book are in one pile.

Workers along the line make sure that there are plenty of signatures in each pocket and that the mechanical arms are working properly. At the end of the line, a worker does a spot check to see that the signatures are in correct order. If they are not, the worker immediately stops the line, while others seek out the cause of the problem.

Stacks of gathered signatures are brought to the section of the bindery where they are sewn together. There is a whole row of sewing machine operators sewing the gathered signatures together. They set the signatures into their machines and press a foot pedal that sends the needle and thread through the backs of the signatures.

The sewn signatures are taken to a press that gives them a powerful, quick squeeze, which compresses the pages and produces a neat package. They now need to be glued off. Some books are not sewn at all, but are only glued.

There is more than one way of gluing a book together, but the most advanced method is the perfect binding way. The perfect binding machine operator watches as the book passes down the line. The paper edges along the back or spine are sliced off and made rough so that there is more surface for the glue. The operator then oversees the application of glue to the back of the book. By the time the book reaches the end of the line, the glue has dried and the back, or binding edge of the signatures are all stuck together.

From this station, all the bound books, sewn and glued or just glued, are brought to the trimmer. The operator controls a machine that cuts around the top, bottom, and front of the book's pages. Now the book has its final size and shape. All the pages are attached, but each is able to turn freely.

The final step in the bindery is to attach the cover to the bound signatures. The covers, or cases, of the book are made from two pieces of heavy cardboard, slightly larger than the pages of the book. A machine glues a covering over the cardboards, joining them into one complete cover.

Printers imprint the title, author, and publisher of the book on the spine and front of the cover. They use a method known as hot die stamping. They stamp with enough pressure to force the image below the surface so that the gold leaf or ink will not be worn away as the book is handled and will be clearly visible on the cloth.

Some books, like this one and the other books in the Industry at Work series, have printed covers. They are not made with the great pressure necessary for imprinting. The printers use the normal pressure, called a "kiss impression," to transfer the image to the plastic-coated cloth of the cover.

Finally, the books are ready to have the covers put on, in a process called casing in. Workers bring the nearly finished books to a machine that covers the endpapers

A worker feeds the endpapers into a machine that will glue them to the signatures.

A worker near the end of the bindery line checks that the book covers are correctly attached.

with glue. A conveyor belt carries the books under a hopper that contains the covers.

As a cover falls onto the pages, clamps on either side paste the endpapers to the cover. The cased-in books travel along and enter another machine, which applies heat and great pressure to the finished books. A metal ridge on the pressure plate forms the joint line on the cover. That is the line along which the covers will bend when the book is opened.

At the end of the finishing line, the bound, cased-in books emerge. Workers collect them and place them in piles on wooden skids for the final step in the bindery process.

Many books also have separate covers, called dust jackets, that are wrapped around the cover. The dust jackets are printed separately and trimmed to the size of the cover. They are placed on the books either by hand or by machine.

With the dust jackets in place, the books are finished. Workers stack them on wooden skids for the last time. They cover the stacks with paper and secure them with metal bands. The books are then shipped by truck or train to warehouses. Here they are kept in big bins, ready to fill the orders that come in from bookstores, libraries, and schools.

FURTHER READING

Arnold, Edmund C. *Ink on Paper. Two.* New York: Harper and Row, 1972.
Liebers, Arthur. *You Can Be a Printer.* New York: Lothrop, Lee and Shepard, 1976.
    A great deal of specific information on careers in printing.
Spector, Marjorie. *Pencil to Press.* New York: Lothrop, Lee and Shepard, 1975. How
    a book is made, from the writing through the printing and binding.
Turnbull, Arthur, and Baird, Russell N. *The Graphics of Communication.* New York:
    Holt, Rinehart and Winston, 1975. An advanced book on printing.

The following booklets on careers in printing are
available from the organizations listed below:

Careers in Graphic Communications
Is Graphic Arts the Career for You?
Answers to Some Questions About Careers in Graphic Communications
Graphic Communications
    Education Council
    Graphic Arts Industry, Inc.
    4615 Forbes Avenue
    Pittsburgh, Pa. 15213

Careers in Printing
    Public Relations Department
    Rochester Institute of Technology
    One Lomb Memorial Drive
    Rochester, N.Y. 14623

Graphic Communications Careers
    Eastman Kodak Company
    Graphic Markets Division
    Rochester, N.Y. 14650

Many large printing plants offer tours of their facilities. For a list of some plants nationwide write for:

USA Plant Visits, 1977–1978. Published by: Superintendent of Documents, U.S. Government Printing Office, Washington, D.C. 20402. (A list of nearly 2000 industrial plants around the country that welcome visitors.)

INDEX

Apprentices, 2, 18

Bindery, 53, 54–61
Binding methods, 57–58
Book covers, 58
Books, 2, 53, 54–61

Camera operators, 27, 29, 33, 36, 38–39, 45
Case, type, 15
Cases. *See* Book covers
Casing in, 58–61
Catalogs, printing product, 7, 10
Chase, 41
Chief executive, 11–13
Color printing, 36–39
    flat color, 36–38
    full color, 38–39
    process color, 38–39
Commercial printers, 2
Composing room, 14–24
Compositors. *See* Typesetters

Computer, 6, 20, 22–24
Customers, 9, 10, 11, 13, 14
Cutting. *See* Slitting
Cutting machine, 55
Cylinder press, 47–49

Dummy, 11, 40
Duplicate plate, 31–33, 41
Dust jackets, 61

Electrotype, 32
Endpapers, 55, 58
Estimators, 9–10

Flats, 45
Folding machine, 54–55
Front office, 7–13
Full color process, 38–39

Galley, 15
Gathering machine, 55–57
Goldenrod, 45

Graphic communication, 6
Gravure, 34–36

Halftone plates, 28–31, 39
Halftone screen, 29, 39
Hot die stamping, 58

Intaglio. *See* Gravure

Job printers, 2
Joint line, 61

Laser beam, 6, 22, 33
Layout artists, 10–11, 40
Letterpress, 33, 49
Linecuts, 25–28, 36–38
Linotype, 17–18

Magazines, 2, 32, 36
Makeready, 46–50, 53
Makeup, 40–41
Monotype, 18–19

Newspapers, 2, 22–24, 32, 36, 41, 45, 49, 53

Offset, 33–34, 49. *See also* Photo-offset lithography

Paste-up, 41–45
Perfect binding, 57–58
Photoengravers, 27–28, 29–31, 38, 39
Photo-offset lithography, 33
Phototypesetters, 19–20
Platemakers, 25–39, 45
Platemaking, 25–39
    color plates, 36–39
    duplicate plates, 31–33
    electrotypes, 32
    gravure, 34–36

    halftone plates, 28–31
    linecuts, 25–28
    offset, 33–34
    stereotypes, 31–32
Platemaking machine, 34
Platen press, 47
Press operators, 38, 39, 46–53
Presses, 14, 38, 46–53, 54
    cylinder, 47–49
    platen, 47
    rotary, 31, 32, 49, 50–53
Pressroom, 46
Pressrun, 50–53
Printers' representatives, 9, 10, 11
Printing plate, 47
Process color. *See* Full color
Progressives, 39
Proofreaders, 24
Proofs, 15, 24, 28, 31, 34, 39
Punched paper tape, 18–19, 20, 24

Repro. *See* Reproduction proof
Reproduction proof, 33, 41
Reps. *See* Printers' representatives
Retoucher, 38
Rotary press, 31, 32, 49, 50–53

Scanner, 22
Shipping room, 53
Signatures, 54, 55–58
Slitting, 54–55
Slug, 18
Stereotype, 31–32
Stick, 15–17
Stripping, 45

Teletype, 22

Trimming, 58
Typesetters, 14–16
Typesetting
    cold metal, 19–20
    hand, 14–17
    hot metal, 17–19
    linotype, 17–18
    newspaper, 22–24

Video Display Terminal, 19, 22

Web presses, 49

CAREER INDEX

Apprentices, 2, 18

Bindery workers, 54–61
Book printers, 2

Camera operators, 27, 29, 33, 36, 38–39, 45
Casing-in workers, 58–61
Chief executive, 11–13
Commercial printers, 2
Compositors. See Typesetters
Cutting machine workers, 55

Estimators, 9–10

Feeding machine operators, 54–55

Gathering machine workers, 55–57

Job printers, 2

Layout artists, 10–11, 40
Linotype operators, 17–18

Magazine printers, 2
Makeup workers, 40–41
Mechanics, 50

Newspaper printers, 2

Paste-up workers, 41–45
Perfect binding machine operators, 57–58
Photoengravers, 27–28, 29–31, 38, 39
Phototypesetters, 19–20
Platemakers, 25–39, 45
Press operators, 38, 39, 46–53
Printers, 1, 2–6, 47, 53, 58
    semiskilled, 2
    skilled, 2, 18
    unskilled, 2, 18
Printers' representatives, 9, 10, 11
Proofreaders, 24

Reps. See Printers' representatives
Retoucher, 38

Sewing machine operators, 57
Strippers, 45

Trimming machine operators, 58
Typesetters, 14–16

Workers, 19, 31, 49, 50, 53, 54, 57, 58, 61